Whack and Yank, Push and Pull

BY KIM THOMPSON

A Little Honey Book

Crabtree Publishing

crabtreebooks.com

Tips for Teachers and Caregivers

This book supports early readers as they decode words to learn facts and gain knowledge about the world.

Before reading, make sure students understand the sound-spelling correspondences shown below as well as the high-frequency words shown on the next page. Introduce the vocabulary words.

During reading, provide feedback and encouragement as students sound out decodable words by blending individual sounds.

After reading, talk about and write about the topic. Share the information on page 16 to help students learn more.

Letters and Sounds

New:

Sound	Spelling
/w/	w
/w/	wh
/y/	y

Review:

Sound	Spelling
short a	a
/b/	b
/k/	c, ck, k
/d/	d
short e	e
/f/	f
/g/	g
/h/	h
short i	i
/j/	j
/l/	l
/m/	m
/n/	n
short o	o
/p/	p
/r/	r
/s/	s
/t/	t
short u	u
/v/	v
/ks/	x

Decodable Words

and, back, gets, hands, in, is, it, not, on, rock, rod, sand, well, wet, whack, whelk, when, whiffs, whip, will, win, wind, yank, yell, yelp, yes, yet

High-Frequency Words

New: find, into, me, play, they, too, with

Review: a, go, I, my, of, the, time, to, water

Vocabulary Words

ball

hair

move

pull

push

waves

"Yes! It is time to play!"
I yell.

"It is time to **move**!"

move

Whiffs of wind **push** me.

They whip my **hair**!

hair
push

I push the **ball**.

I will whack it to win!

I **pull** a rock with my hands.

In a well of water, I find a whelk!

When my rod gets a yank, I yelp.

I pull and pull!

Wet **waves** push on the sand.

They pull back into the water.

I move sand, too.

It is not time to go yet!

Build Background Knowledge

It takes force to make something move. You can use force to make something move by pushing it or pulling it. Pushing something moves it away. A stronger push will move something away faster and farther. Pulling something brings it closer. A harder pull will make it come faster and closer. How are pushes and pulls making things move at the beach in this book? Think about what you did today. How did you make things move by pushing and pulling?

Written by: Kim Thompson
Designed by: Rhea Magaro
Series Development: James Earley
Educational Consultant: Marie Lemke, M.Ed.

Photographs: All images from Shutterstock

Crabtree Publishing

crabtreebooks.com 800-387-7650

Printed in China/012024/FE20231222

Published in Canada
Crabtree Publishing
616 Welland Ave.
St. Catharines, Ontario
L2M 5V6

Published in the United States
Crabtree Publishing
347 Fifth Ave
Suite 1402-145
New York, NY 10016

Library and Archives Canada Cataloguing in Publication
Available at Library and Archives Canada

Library of Congress Cataloging-in-Publication Data
Available at the Library of Congress

Hardcover: 978-1-0398-4435-3
Paperback: 978-1-0398-4516-9
Ebook (pdf): 978-1-0398-4593-0
Epub: 978-1-0398-4663-0
Read-Along: 978-1-0398-4733-0
Audio: 978-1-0398-4803-0